**Wallisson Oliveira Ferreira**

# Removal of copper ions using baru bark and pulp as an adsorbent

Wallisson Oliveira Ferreira

# Removal of copper ions using baru bark and pulp as an adsorbent

## An analysis of the adsorption capacity of Baru (Dipteryx alata) as an adsorbent for copper ions in aqueous systems

ScienciaScripts

**Imprint**



Cover image: www.ingimage.com

This book is a translation from the original published under ISBN 978-613-9-63189-6.

Publisher:
Sciencia Scripts
is a trademark of
Dodo Books Indian Ocean Ltd. and OmniScriptum S.R.L publishing group

120 High Road, East Finchley, London, N2 9ED, United Kingdom
Str. Armeneasca 28/1, office 1, Chisinau MD-2012, Republic of Moldova, Europe
Printed at: see last page
**ISBN: 978-620-7-77374-9**

# SUMMARY

*"I was taught that the road to progress is neither quick nor easy"*

*(Marie Curie)*

## SUMMARY

The use of water resources has intensified with economic and population growth. Industrial activity and the excessive use of agrochemicals are the main culprits in contaminating water bodies. There are various methods of treating industrial effluents with the aim of minimizing the environmental impact caused by these contaminants, especially metal ions, which even in low concentrations are extremely harmful to living beings. With the need to develop increasingly viable and efficient methods, adsorption studies using natural materials have been highlighted, which can even be obtained from by-products of agro-industrial processes, making the process more economically viable. This study evaluated the potential of the bark and pulp of the Baru species *Dipteryx alata* as a natural adsorbent for removing copper ions Cu (II) from aqueous systems and also characterized the material in terms of its composition using the analytical techniques of Infrared Spectroscopy (FTIR) and Thermogravimetric Analysis (TGA). The adsorption parameter optimization tests to determine the best removal conditions were carried out using a univariate method, the variables studied were: pH (3, 5, 7 and 8); agitation time (5, 15, 20, 30, 45 and 60 min) and adsorbent mass (25, 50, 100, 150 and 200 mg). Removal was carried out with standard solutions of Cu (II) at a concentration of 10 mg $L^{-1}$ , after stirring the supernatant was filtered and analyzed by Flame Atomic Absorption Spectroscopy (FAAS). The highest Cu (II) removal capacity was observed for the following conditions: solution pH 7.0; stirring time of 20 minutes and adsorbent mass of 25 mg. The copper removal results fitted the Langmuir and Freundlich models, with close linear correlation coefficients. The Langmuir model obtained $Q_{max}$ (mg $g^{-1}$ ) of 22.12; b (L $mg^{-1}$ ) of 0.1471 and a correlation coefficient of 0.9782, and for the Freundlich model we obtained Kf (L $mg^{-1}$ ) of 3.645 and a constant 1/n of 0.508 indicating that the isotherm is favorable and a correlation coefficient of 0.9322, suggesting the occurrence of monolayer adsorption as well as heterogeneous conditions on the surface. In view of the results obtained, it can be concluded that the material showed potential for removing Cu (II) ions from aqueous systems, indicating the possibility of using Baru as an alternative material for chemical remediation and effluent polishing.

Keywords: Baru. Adsorption. Copper.

# 1 INTRODUCTION

Water is a natural resource of fundamental importance for the existence of life and for the conservation of biodiversity. It is a precious resource of inestimable value that must be conserved and protected. The use of water resources has intensified with economic growth and population increase, producing numerous alterations to the hydrological cycle and water quality (BACCI, PATACA, 2008).

All forms of consumption of water resources generate waste at the end of their utilization processes. This waste contains organic and/or inorganic contaminants that can cause significant changes to ecosystems and consequently affect people's quality of life (THEODORO, 2010).

According to NAGATA and collaborators (2010), society's great challenge is to make everyone aware of the need for environmental preservation, minimize the effects of human decisions on the environment and at the same time promote development, in addition to the need to stop wasting water and develop more efficient methods of saving, treating and reusing it (BRASIL, 2005).

The industrial and agricultural sectors are the biggest consumers of water and also largely responsible for its waste, pollution and contamination, mainly due to the lack of development of more efficient and economical industrial processes, inefficient or incorrect treatment of effluent, incorrect disposal of waste and excessive use of pesticides and fertilizers in agricultural production (BRASIL, 2005).

Among the most varied types of contaminants, toxic metals represent a major threat to human health, since when they come into contact with organic matter they form highly stable compounds that can accumulate in tissues and become potentially cytotoxic and carcinogenic, since the human body is unable to process and/or eliminate them (THEODORO et al, 2011).

These contaminants end up being discharged into water bodies due to the difficulties of completely removing metals from effluents using conventional treatment methods, which are mainly low removal efficiency, low selectivity and high application costs.

In addition, the methods currently available are only applied to effluents with low metal concentrations (MÓDENES et al., 2013).

According to TERA AMBIENTAL (2014), in order to improve the efficiency of an effluent treatment process, the composition of the effluent must first be analyzed to determine which type of treatment is best suited to meet the specifications of local legislation. The main physico-chemical methods used in industry are: chemical precipitation, adsorption with activated carbon, electrocoagulation, electrodeposition, reverse osmosis and evaporative recovery (AYDIN, BULUT, YERLIKAYA, 2008).

Brazil's socio-economic situation makes it essential to invest in the development of low-cost, efficient alternative technologies for wastewater treatment, in which natural adsorbents have stood out (SABEI, BASSETTI, 2013).

According to MOSQUETTA and collaborators (2011), natural adsorbents are materials that are not produced synthetically and that have adsorptive properties for organic and/or inorganic chemical species, mainly metal ions and coloring pigments. Among these are lignocellulosic materials, organic materials whose structure includes a wide variety of functional groups, such as carboxylic acids, phenols, amines and amides, which can retain heavy metals by ion exchange and/or electrostatic interaction. In addition, they are rich in cellulose and lignin, serving to adsorb metals by hydrophobic interaction and van der Waals forces (AMAVISCA, 2012).

Lignocellulosic adsorbents have advantages over conventional methods because they are low-cost, they can be used with materials that are by-products or residues of some agricultural production processes, they are highly efficient at removal, they minimize the use of chemical products and the metal can be recovered (MOSQUETTA et al, 2011).

According to SOUZA (2012), Baru (*Dipteryx alata*) is a tree species native to the Brazilian cerrado region, mainly in Goiás, Minas Gerais and Mato Grosso do Sul. The fruit consists of a yellow pulp (mesocarp) and a brown kernel with high nutritional value in oils, fibers and especially proteins (CRUZ et al., 2011).

This work aims to investigate the potential of Baru (*Dipteryx alata*) bark and pulp as a natural adsorbent for removing Cu (II) ions from aqueous solutions, as well as optimizing the system variables to determine the best removal conditions and characterizing the material using Infrared Spectroscopy (FTIR) and Thermogravimetric Analysis (TGA).

# 2 LITERATURE REVIEW

## 2.1 Environmental education and sustainable development

The economic development model often seeks growth without worrying about the depletion of natural resources and consequently about pollution and the destruction of biodiversity (COSTA, IGNACIO, 2011).

The population needs to be sensitized to the problems that anthropogenic actions are causing and must work together with government bodies to preserve the environment and make sustainable use of resources. To do this, it is necessary to spread the ideas of environmental education in terms of preserving and restoring the environment and controlling polluting activities (MARCATTO, 2002).

*Environmental education must be guided by a systemic approach, capable of integrating the multiple aspects of contemporary environmental issues [...] Even more than a systemic approach, environmental education requires the perspective of complexity, which implies that different levels of reality interact in the world (objective, physical, abstract, cultural, affective...) and that different perspectives are constructed as a result of different cultures and individual and collective trajectories (Ministry of the Environment, 2014).*

In 2004, the United Nations Organization began developing the Decade for Sustainable Development (2005 to 2014), designating UNESCO as the international leader and coordinator. The aim was to integrate the values inherent in sustainable development in order to foster changes in behavior that would create a sustainable and egalitarian society (©UNESCO, 2005).

At the ECO-92 conference, the Chamber of Deputies (1995) established the objective of creating the Brazilian Agenda 21, finalized in 2004, which has as its general principles environmental protection and the use of natural resources and the elaboration of sustainable development policies (Ministry of the Environment, 2004).

Sustainable development can be defined as development that meets the needs of current generations without compromising the needs of future generations to meet their own. The difficulty of establishing economic development without harming the environment is one of the greatest difficulties, given that, from today's perspective, in order for the economy to grow there must necessarily be some damage to the natural environment, even if it is minimal (MESQUITA et al, 2014).

## 2.2 Water

Water is a limited natural resource that is essential for the survival of all species, for personal hygiene, domestic use, industrial production, irrigation in agriculture and various other activities.

This natural resource can be found in three different phases: solid, liquid and gas. In the solid phase, it is found in glaciers, mainly in the region of the poles, and makes up approximately 2% of the total volume of water on the planet; in the gaseous phase, water is present throughout the atmosphere in the form of vapor; the liquid phase represents approximately 98% of its entire volume on earth, 97% of which is made up of the oceans and around 1% of which is made up of continental waters, which can be on the surface or underground. Surface water corresponds to rivers, lakes and ponds, and underground water to groundwater (THEODORO, 2010).

The supply of drinking water and the quality with which this resource reaches the population for use has become a worldwide concern. The fact that long periods of drought are largely due to the imbalance caused by global warming has made it the subject of studies, dissertations, theses and conferences all over the world (BIANCHI, 2011).

According to the National Water Agency (2014), since 2012 there has been a gradual and intense reduction in rainfall in some regions of the country and consequently a reduction in the level of the main reservoirs. The threat of a water crisis and scarcity of drinking water is generated mainly by overconsumption, but also by geographical factors, for example in Brazil, which is home to a large volume of water reserves, but is already affected in some regions by the crisis due to non-uniform availability (PROSAB, 2009).

*The insufficient supply of water for various uses has a direct effect on the daily lives of the population and the country's economy and has demanded great attention in the search for alternatives to deal with the problems: emergency measures and structural measures, as well as non-structural measures related to improving the management of water resources in Brazil. Supporting and improving water reuse techniques, reducing waste by the different user sectors (in irrigation, industry, distribution and residential consumption, for example), as well as implementing actions to conserve water sources are among the measures that should be prioritized and promoted (Agência Nacional de Aguas, 2014).*

Figure 1 shows the distribution of water resources, surface area and population as a percentage of Brazil's total.

Figure 1 - Distribution of water resources, surface area and population in Brazil.

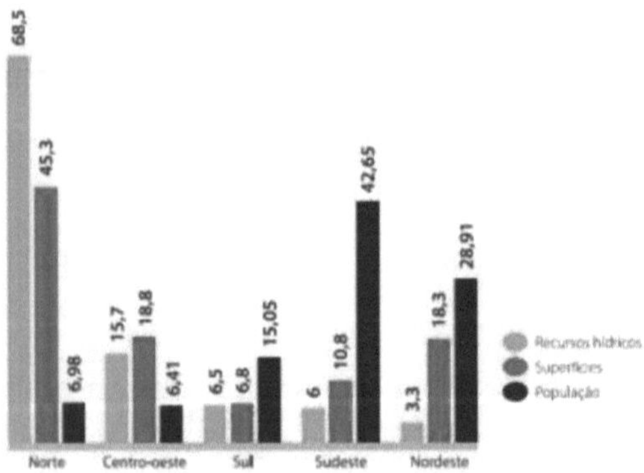


Source: Ministry of the Environment, Brazil (2005)

According to RIBEIRO et al. (2012), water quality is another factor that limits the use of this resource. In addition to the lack of basic sanitation and treated water in some regions, inorganic contaminants have gained visibility through the study of contamination by toxic metals. New methods of removing these contaminants are being developed to avoid problems for the health of living beings (AMALINEI et al., 2012).

### 2.3 Toxic metals and contamination risks

According to PEDROTTI and MISTURA (2010), the main sources of metal emissions are: ash and particulate emissions from the burning of fossil fuels, the disposal of industrial products, especially in the metal, steel, pulp and paper, chemical and petrochemical industries, the abusive use of pesticides and fertilizers in agriculture, domestic effluents and the disposal of sewage sludge. These metals can contaminate soil, air and aquatic environments and when they reach high concentrations they cause toxicity problems in plants, animals and humans (STEFFEN, STEFFEN, ANTONIOLLI, 2011).

The main threats to human health from toxic metals are associated with exposure to lead, cadmium, copper, mercury and arsenic. These metals can cause lung, kidney, psychological, neurological, gastrointestinal and cardiovascular damage and may even be associated with the development of lung cancer in smokers and workers in some mining companies (SEMEDO, 2014).

In order to regulate the conditions, parameters, standards and guidelines for managing the discharge of effluents into receiving bodies of water, bodies such as CONAMA draw up resolutions to try to prevent or reduce contamination (BRASIL, 2011).

Table 1 shows some standards and conditions for effluent discharge:

Table 1 - Discharge standards according to CONAMA RESOLUTION No. 430 of May 13, 2011

| **Inorganic parameters** | **Maximum values** |
| --- | --- |
| Total arsenic | 0.5 mg $L^{-1}$ |
| Total cadmium | 0.2 mg $L^{-1}$ |
| Total lead | 0.5 mg $L^{-1}$ |
| **Dissolved copper** | **1.0 mg $L^{-1}$** |
| Total tin | 4.0 mg $L^{-1}$ |

| Total mercury | 0.01 mg $L^{-1}$ |
|---|---|

Source: CONAMA RESOLUTION No. 430 (2011)

### 2.4 Copper

Copper is a metal found in all the rocks of the earth's crust. It has a great ability to interact chemically with mineral and organic components and is considered the most immobile of metals, accumulating on the surface in the form of oxides, sulphates, carbonates and hydroxides (KABATA-PENDIAS, PENDIAS, 1992).

Copper is one of the world's most widely used raw materials in a variety of applications. Its widespread use is favored by its high availability and recyclability, as well as its metallic properties which make it an excellent conductor of heat and electricity. In addition to industrial applications, such as the manufacture of bronze, wires for the electricity grid, various vehicle components and electroplating processes combined with leveling to prevent corrosion, due to its high added value copper is an accurate indicator of the world economy (RODRIGUES, SILVA, GUERRA, 2012).

Copper sulphate is one of the most widely used chemical products in agriculture and livestock farming to treat various problems such as: problems with copper deficiency in the soil, fungicide and bactericide to protect valuable crops, and animal health and welfare (ELUMA S/A, 2010).

In 2011, around 16 million tons of copper were produced worldwide, with Chile being the largest producer, as shown in figure 2:

Figure 2 - Major copper producers

**World Copper Production**

Data in thousands of metric tons of copper content World production: 16,200 tons (2011)

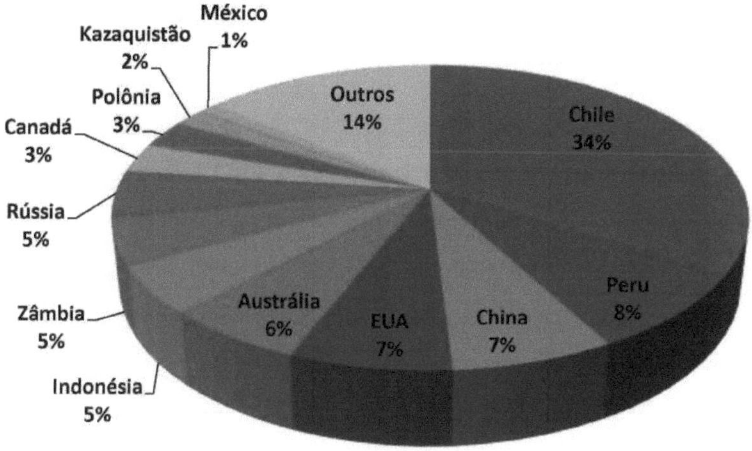


Source: www.rankia.com (adapted) (2011)

The liquid effluents generated by these industrial processes contain copper and other metals in the form of dissolved salts and organic complexes. In order to reduce the negative impacts generated on the environment, various effluent treatment methods are available for use. Among these methods, electrocoagulation stands out due to its versatility and environmental compatibility (THEODORO et al, 2011).

In the electrocoagulation process, the contaminants are subjected to strong electric fields, causing oxidation-reduction reactions and being brought to less reactive, insoluble and more stable chemical states, forming flakes. Water is also electrolyzed in a parallel reaction, producing small bubbles of oxygen at the anode and hydrogen at the cathode. These bubbles attract the flocculated particles, causing them to float to the surface and then be removed (MENESES et al, 2012).

Some metals are essential for vital metabolic processes in organisms, but others such as copper, even in relatively low concentrations, are responsible for risks of damage to the skin, lungs and the development of cancer (AMALINEI, 2012).

When metals accumulate in the body, they are generally not destroyed by microbial processes and human metabolism. These contaminants act directly on the enzymatic activity of living beings, blocking biological activities and inactivating enzymes due to the formation of bonds between the metal and some functional groups of proteins, which can cause irreversible damage to various organisms (SILVA, 2014).

## 2.5 Adsorption

According to UCUN, AKSAKAL and YILDIZ (2009), adsorption is defined as the accumulation of molecules on the external surface of a solid adsorbent material through contact in solution, until equilibrium is reached. The adsorption mechanism is the result of the formation of complexes between functional groups (amines, hydroxyls and carboxyls) on the surface of the adsorbent and the chemical species in solution (AMAVISCA, 2012).

According to SILVA (2014), the adsorption process depends on several factors such as: the nature of the adsorbent, the adsorbate and the operating conditions. The characteristics of the adsorbent include surface area, density, functional groups present on the surface and hydrophobicity of the material. On the other hand, the nature of the adsorbate depends on the polarity, size of the molecule, solubility, acidity or basicity, and the operating conditions of the adsorption process, which mainly include temperature, pH of the solvent and its nature (COONEY, 1999; RUTHVEN, 1984; MCKAY, 1996).

Generally, adsorption studies are carried out on monometallic systems. Multimetallic systems have been studied very little, but it is known that the competitive interaction between metals in solution is a factor that affects the degree of removal (GAVRILESCU, 2004).

Treatments using ion exchange resins and adsorption methods using silica, alumina and activated carbon are used in industry to control pollution, but these are expensive methods. Alternative treatments using low-cost natural biological materials have been studied (UCUN, AKSAKAL, YILDIZ, 2009).

## 2.6 Natural adsorbents

Natural adsorbents are materials found in nature with a high capacity for adsorption and removal of toxic substances from aqueous systems. The adsorption or biosorption capacity of these adsorbents can be caused by physical adsorption in the layered structure, ion exchange processes and the formation of hydrogen bonds, often due to multiple interaction factors (SINEVA, 2014).

Natural adsorbents can be classified as: biosorbents, mineral adsorbents, lignocellulosic adsorbents. Various natural materials with adsorption potential can be obtained from waste generated in industrial processes, making the use of natural adsorbents more economically viable than conventional methods, as well as helping to reduce the cost of waste disposal (PARAB, 2010).

Biosorption should be applied in a tertiary stage of treatment, also called the polishing stage, which takes place after the preliminary, primary and secondary stages of wastewater treatment and has the purpose of achieving additional pollutant removals before the effluent is discharged into a receiving body.

The main advantages of this type of renewable treatment are: the low cost of obtaining it, the high rate of renewal in nature, the potential for removing metals and the possibility of recovering the contaminant, either by incinerating the biomass or by a desorption process (MÓDENES et al., 2013).

A natural material to be used as a biosorbent must have the following characteristics: it must be low cost; it must have a particle size, shape and mechanical strength suitable for use in bioreactors under continuous flow conditions; metal capture must be efficient and fast; separation of the biosorbent from the solution must be fast, efficient and cheap and it must have high selectivity (SILVA, 2014).

Figure 3 shows some biosorbents tested for the removal of metal ions:

Figure 3 - Biosorbents tested for the removal of different metal ions

| **Metal** | **Copper (Cu)** | **Lead (Pb)** | **Chromium (Cr)** | Zinc (Zn) |
|---|---|---|---|---|
| **Bk\| ssf] rτentes Aviliados** | Corn leaf | Green Algae | Green Algae | Olive Oil Residues |
| | Biomass waste | *Candida albicans* (fungus) | Japanese Cedar Bark | Yeasts |
| | Orange peel | *Rhodo tonila ghitinis* (levedura) | Egg shells | *Pseudomona cepacia* (bacteria) |

Source: Adapted from MUDHOO et al. (2012)

## 2.7 Lignocellulosic adsorbents

According to DERMIBAS (2008), lignocellulosic adsorbents are materials with a porous surface capable of adsorbing molecules on their surface. They stand out for having lignin (Figure 4) and cellulose in their structure, macromolecules with a wide variety of adsorption sites. Some examples of low-cost lignocellulosic adsorbents are: Baru (*Dipteryx alata*) bark and pulp, Romanian pine (*Pinus sylvestris L.*) bark, lentil (*Lens culinaris Medik)* shells, wheat (*Triticum durum Desf*) and rice (*Oryza sativa L.*) (AYDIN, BULUT, YERLIKAYA, 2008; AMALINEI, 2012).

Figure 4 - Structure of the lignin molecule

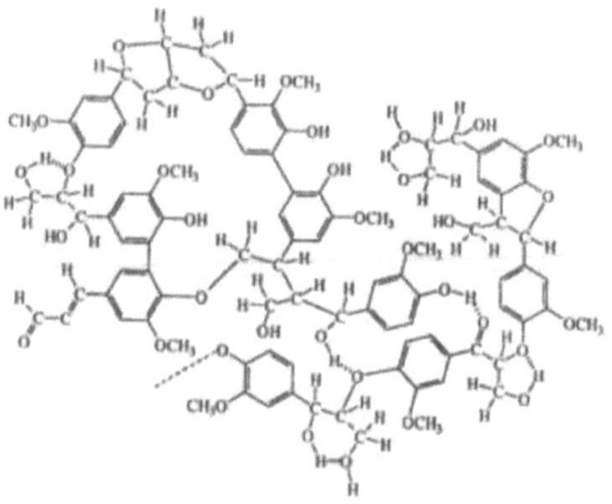

Source: SOUZA (2006)

Cellulose is a linear homopolysaccharide made up solely of glucose molecules joined together by $\beta$ (1→4) glycosidic bonds, which result from the loss of a water molecule. Lignin, Figure 5, is a macromolecule synthesized by the radical route from three precursor *p-hydroxy* cinnamyl alcohols: *p-coumaryl*, coniferyl and sinapyl.

Figure 5 - Molecular structure of a cellulose chain

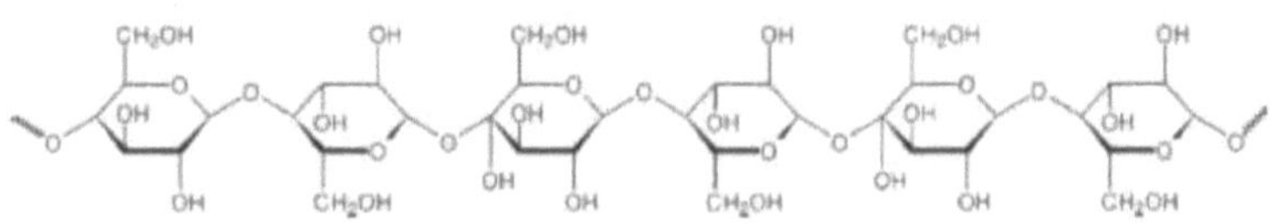

Source: SANTOS et al. (2001)

### 2.7.1Baru

The Brazilian cerrado is a savannah that deserves to be preserved and valued for its great and rich biodiversity. One of the native species of this biome is the barueiro (*Dipteryx alata*), a tree of the Leguminosae family, abundant in the states of Mato Grosso, Mato Grosso do Sul, Goiàs and Minas Gerais (CRUZ et al., 2011).

According to CARRAZZA and D'AVILA (2010), the barueiro is a large tree, reaching 25 meters in height and 70 cm in diameter. It bears its first fruit at around 6 years of age, and this period varies greatly depending on the soil and water conditions. Fruiting is seasonal and varies according to region and soil type. The fruit is formed by a spongy mass that covers an ovoid kernel with a brown shell. The kernel is a significant source of lipids, protein (30%) and, consequently, calories, as well as dietary fiber (MOSQUETTA et al, 2011).

The total protein fraction of legume seeds is a complex mixture of albumin, globulin, prolamins and glutelins, and as the largest fraction, globulin is often responsible for the physical and nutritional properties of the products. Baru seeds have a higher protein content than common beans (*Phaseolus vulgaris L.),* chickpeas (*Cicer aretinum L.*) and lentils (*Lens culinaris Med).* The high protein content of Baru seeds has attracted interest in the development of protein-based products (CRUZ et al., 2011). Figure 6 shows Baru fruits of the *Dypterix alata* species:

Figura 6 - Baru tree and Baru fruit (bark and kernels)

Source: Institute of Health, Population and Nature (2010) and Ministry of Agrarian

Development (2014)

## 2.8 Isotherms

Adsorption isotherms are mathematical models that describe adsorption processes based on the association between the mass law equation and the electrostatic effects caused by variations in the charges of the adsorbent surfaces, evaluating the adsorption capacity and the strength with which the metals bind to the adsorbent (GAVRILESCU, 2004). These models can be built empirically or obtained from mathematical modeling.

There are several types of adsorption isotherms which are represented by four classes according to their initial shape: S, L, H and C (GILLES et al, 1960). Figure 7 shows these classifications:

Figura 7 - Classification of Adsorption Isotherms

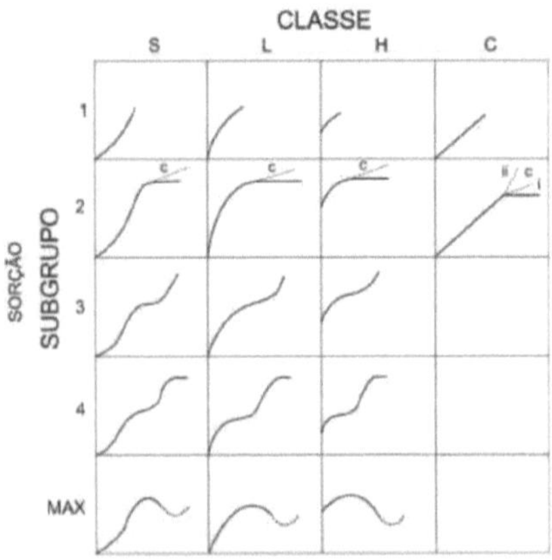


Source: GILLES et al. (1960)

- S (sigmoidal): S-shaped, concave and followed by an inflection point.
- L (Langmuir): These are the most common, offer downward concavity due to the decreased availability of active sites and represent adsorption in monolayers.
- H ("high affinity"): These correspond to extremely strong adsorptions in the region of low concentrations, and are highly inclined in these regions, implying strong preferential adsorption of the solute.

- C ("Constant partition"): Common in microporous adsorbents, they initially have a linear portion, indicating constant partitioning of the solute between the solution and the surface of the adsorbent.

According to the shape of the isotherm curve, it is possible to obtain information about the adsorption process. It can be said that when the isotherm is linear, passing through the origin, the amount adsorbed is proportional to the concentration of the fluid. Isotherms with a downward concavity are said to be favorable, as large amounts of adsorbate can be obtained at low solute concentrations (DO, 1998).

The most widely used models are those of Freundlich and Langmuir. The Langmuir equation provides a quantitative parameter, related to the maximum adsorption capacity, and a qualitative one, which expresses the binding energy (LINHARES et al, 2008).

The mathematically deduced equation has a general formula, as follows

Equation 1:

$$Q_e = \frac{b * Q_{max} * C_e}{1+(b * C_e)} \qquad (1)$$

Linearizing this equation, we get Equation 2:

$$\frac{1}{Q_e} = \frac{1}{Q_{max}} + \frac{1}{b * Q_{max}} * \frac{1}{C_e} \qquad (2)$$

Where $Q_e$ is the amount of analyte adsorbed at equilibrium (mg $g^{-1}$ ); *b* and $Q_{max}$ are Langmuir parameters related to the adsorbent- adsorbate interaction forces and the Maximum Adsorption Capacity, respectively, *and* $C_e$ is the concentration of the analyte at equilibrium (mg $L^{-1}$ ) (TAGLIAFERRO et al, 2011).

The dimensionless constant $R_L$ can be used to predict the shape of the adsorption isotherm according to the Langmuir model, indicating whether adsorption is favorable or unfavorable. This constant, also called the degree of equilibrium, indicates the degree of adsorption and is defined in Equation 3:

$$R_L = \frac{1}{1+(b * C_e)} \qquad (3)$$

Table 2 shows the relationship between the type of isotherm and the RL value:

**Table 2 - Limits of RL values for Langmuir isotherm**

| **RL values** | **Type of Isotherm** |
|---|---|
| RL > 1 | Unfavorable |
| RL = 1 | Linear |
| 0 < RL <1 | Favorable |
| RL = 0 | Irreversible |

The Freundlich isotherm is an empirical model that describes experimental results of adsorption in aqueous systems and is restricted to the monolayer adsorption model (LINHARES et al, 2008).

The Freundlich isotherm can be described according to Equation 4:

$$Q_e = K_F * C_e^{1/n} \tag{4}$$

This equation can be expressed in linearized form, giving Equation 5:

$$\log Q_e = \log K_F + \frac{1}{n}\log C_e \tag{5}$$

Where $Q_e$ is the amount of the analyte adsorbed at equilibrium (mg $g^{-1}$), $K_F$ is the Freundlich constant; $n$ is the dimensionless constant related to the intensity of adsorption (linearity factor) and $C_e$ represents the concentration of the analyte species at equilibrium (mg $L^{-1}$). $K_F$ and n are empirical Freundlich constants obtained by linearizing equation 4 (FREUNDLICH, 1906).

The constant 1/n has a value between 0 and 1 and is related to the heterogeneity of the surface. The closer it is to zero, the more heterogeneous the surface. The Kf constant is an approximate measure of the adsorption capacity of the adsorbent. The higher its value, the greater the adsorption capacity.

# 3 OBJECTIVES

## 3.1 General objective

To evaluate the potential of crushed Baru (*Dipteryx alata*) bark and pulp as a natural adsorbent for removing Cu (II) ions from water and to characterize the material using FTIR and TGA techniques.

## 3.2 Specific objectives

- Evaluate and determine the parameters for best adsorption: pH of the solution, mass of the adsorbent and stirring time, using the univariate method;
- Determine the maximum adsorptive capacity (MAC) of the material in the removal of Cu (II) ions by constructing adsorption isotherms.
- Characterize the material by Infrared Spectroscopy (FTIR) and Thermogravimetric Analysis (TGA);

# 4 EXPERIMENTAL PART

## 4.1 Necessary infrastructure

- TE-42 Magnetic Shaker (Tecnal, Piracicaba, SP)
- TE-650 Knife Mill (Tecnal, Piracicaba, SP)
- Electronic Analytical Balance FA 2104N, (Bioprecisa, Belo Horizonte, MG)
- Flame Atomic Absorption Spectrometer - FAAS Analyst 400 (Perkin Elmer, Massachusetts, USA)
- Tyler Sieves (Bertel - Ind. Metalùrgica Ltda , Caieiras, SP)
- Greenhouse with air circulation model MA-035, (Marconi, Piracicaba, SP)
- Milli-Q Millipore Simplicity 185 water purifier/deionizer system (Millisul Ind.Com. Quim. Biotec. Ltda, Porto Alegre, RS)
- MA522 Portable Microprocessor pH Meter (Marconi, Piracicaba, SP)
- Quanty JP-42 Filter Paper
- Infrared Spectrometer - FTIR 1605 (Perkin Elmer, Massachusetts, USA)
- Thermogravimetric Analyzer - TGA 2950 (TA Instruments, New Castle, USA)
- Cu (II) standard 1000 mg $L^{-1}$
- NaOH solution 0.3 mol $L^{-1}$
- HNO3 solution 0.3 mol $L^{-1}$
- Glassware in general.

## 4.2 Methodology

### 4.2.1Obtaining and preparing Baru (*Dipteryx alata*)

The Baru fruits were obtained from the courtyard of the State University of Goiás Campus Anàpolis de Ciências Exatas e Tecnológicas Henrique Santilo, in October 2014. The peels and pulp were separated from the kernels using a mechanical method, then dried at a temperature of 34°C for 48 hours in an air-circulating oven model MA-035 (Marconi, Piracicaba, SP) and ground in a TE-650 knife mill (Tecnal, Piracicaba, SP). To obtain the adsorbent in the desired particle size, it was separated on Tyler sieves (Bertel - Ind. Metalalùrgica Ltda, Caieiras, SP) with particle sizes of $115 \leq G \leq 100$ mesh. The material was stored in polyethylene bottles at room temperature.

### 4.2.2Study of variables in adsorption capacity

To determine the most favorable conditions for using Baru as an adsorbent for copper ions, the stirring time (minutes) between the adsorbent and the solution containing the analyte, the mass of the adsorbent (milligrams) and the pH of the solution were evaluated using the univariate method.

The experiments were carried out in triplicate, with a standard Cu (II) solution with a concentration of 10 mg $L^{-1}$ prepared from the analytical grade standard with a concentration of 1000 mg $L^{-1}$ (Dinàmica Quimica Contemporànea Ltda, Diadema-SP).

The pH was adjusted using a MA522 pH meter with solutions of sodium hydroxide (NaOH) and nitric acid ($HNO_3$), both at a concentration of 0.3 mol $L^{-1}$ . The Cu(II) solutions were stirred with the respective mass of adsorbent at room temperature in polyethylene flasks and a TE-42 Shaker mechanical stirrer (Tecnal, Piracicaba, SP). After shaking, the solutions were filtered through a simple filtration system using Quanty JP-42 filter paper. The supernatant was analyzed by Flame Atomic Absorption Spectroscopy (FAAS), Perkin Elmer, Analyst 400 (Massachusetts, USA) for the quantification of Cu (II) ions.

#### 4.2.2.1.1 Study of pH variation

In order to determine the best pH for removing copper ions using Baru as an adsorbent, studies were carried out with successive pH variations, keeping the mass of adsorbent and stirring time constant. The pH values studied were: 3, 5, 7 and 8.

25.0 mg of adsorbent and 15.0 ml of standard solution calibrated to the desired pH were placed in each flask, and the solution was shaken for 20 minutes at room temperature.

#### 4.2.2.1.2 Study of the variation in stirring time

In order to determine the best time to stir the standard solution with the adsorbent to remove Cu (II) ions, studies were carried out with successive time variations, keeping the mass of adsorbent and pH equal to 7. The time values studied were: 10, 15, 20, 30, 45 and 60 minutes.

#### 4.2.2.1.3 Study of adsorbent mass variation

To determine the best adsorbent mass for removing Cu (II) ions, the mass was varied, keeping the stirring time equal to 20 minutes and the pH equal to 7, according to the results of the optimization. The adsorbent mass values studied were: 25, 50, 100, 150 and 200 mg.

#### 4.2.2.1.4 Characterization of Baru (*Dipteryx alata*)

The characterization of the crushed material in defined granulometry was carried out at the State University of Goiás Campus Anàpolis de Ciências Extas e Tecnológicas Henrique Santilo. Infrared Spectroscopy (FTIR) analysis was carried out to determine the presence of functional groups associated with adsorption phenomena, Thermogravimetric Analysis (TGA) to evaluate the loss of mass of the material in relation to temperature variation in an environment with a controlled atmosphere.

#### 4.2.3.1 Infrared Spectroscopy (FTIR)

The baru bark and pulp "in natura" and after adsorption were characterized by Infrared Spectroscopy, using the FTIR-1605 spectrophotometer model (Perkin Elmer, Massachusetts, USA) to identify the functional groups present in the material. For the experiment, the sample was mixed with KBr in a ratio of 1:100 (m/m) to produce a tablet by pressing at 9 tons of pressure. The spectral range varied from 4000 to 400 $cm^{-1}$ .

#### 4.2.3.2 Thermogravimetric Analysis (TGA)

A study to characterize the "in natura" adsorbent material was carried out on the

Thermogravimetric Analyzer model TGA 2950 (TA Instruments, New Castle, USA) at the State University of Goiás. The experiment was carried out in an inert atmosphere of nitrogen gas ($N_2$ ), subjecting 7.047 mg of the material to a heat flow of 10°C/min with an initial temperature of 25°C and a final temperature of 900°C. The purpose of this analysis is to evaluate the loss or addition of mass to the sample at varying temperatures.

# 5 RESULTS AND DISCUSSION

## 5.1 Study of variables in adsorption capacity

### 5.1.1Study of pH variation

pH is one of the most important variables in wastewater treatment and in the adsorption process, directly influencing the concentration of free ions in solution and their electrostatic interaction with the adsorption sites of the adsorbent.

The low adsorption of the metal ion at acidic pH is due to the presence of high concentrations of the proton, causing competition with Cu(II) ions for adsorption sites. At alkaline pH (pH > 7.0) a decrease in the percentage of removal was observed, which can be explained by the high concentration of OH- which can compete with the metal for the sites and/or precipitate it in the form of hydroxide (HASSAN, ABDEL-MOHSEN, ELHADIBY, 2014).

The effect of pH on the adsorption of Cu (II) ions was evaluated by varying the following pH values: 3, 5, 7 and 8. The optimum working pH chosen was 7.0, where good removal values were observed with Baru bark and pulp. It is also interesting to work in this range, so that at the end of treatment the pH of the treated effluent is already between the minimum and maximum values (5.0 to 9.0) for the emission of effluents permitted by Resolution No. 357 of 17/03/2005. The results obtained in this study are shown below in Figure 8.

Figure 8 - Effect of varying pH as a function of Cu (II) ion removal using

Baru

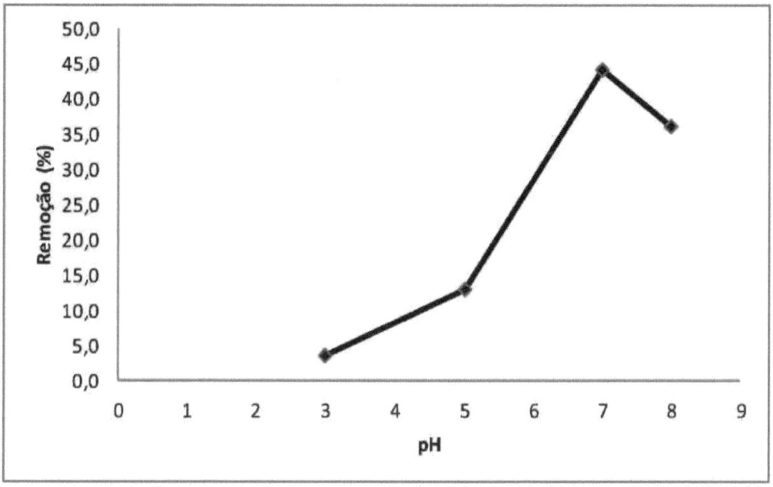


Conditions: m = 25.0 mg, stirring time = 20 min, solution volume = 15.0 mL and [Cu (II)] = 10 mg $L^{-1}$

### 5.1.2 Study of the variation in adsorption time

Knowing the optimum time for the interaction between Cu (II) and the natural adsorbent is an important parameter to be studied in order to assess the retention time required for adsorption to occur with satisfactory values.

The effect of stirring time on the adsorption of Cu (II) ions was studied at 10, 15, 20, 30, 45 and 60 min, as shown in Figure 9. The amount of Cu (II) ions retained by the adsorbent material increased considerably up to 20 min, showing an increase in the percentage of removal. However, between 30 and 60 min there was a decrease in the percentage between the amounts of ions adsorbed per unit mass of the adsorbent.

The time of 20 minutes was chosen due to the good results presented in adsorption, and it also provides speed and agility in laboratory analysis practices. Values of less than 20 minutes are not recommended in order to guarantee the efficiency of the process, which at equilibrium corresponds to the maximum adsorbed capacity for the specific variables. In the values after 20 minutes, removal decreased, possibly due to saturation of the material's active sites, desorption due to agitation and also due to the variation in the solution's pH during the agitation period.

Figure 9 - Effect of varying the stirring time on the removal of Cu (II) ions using Baru

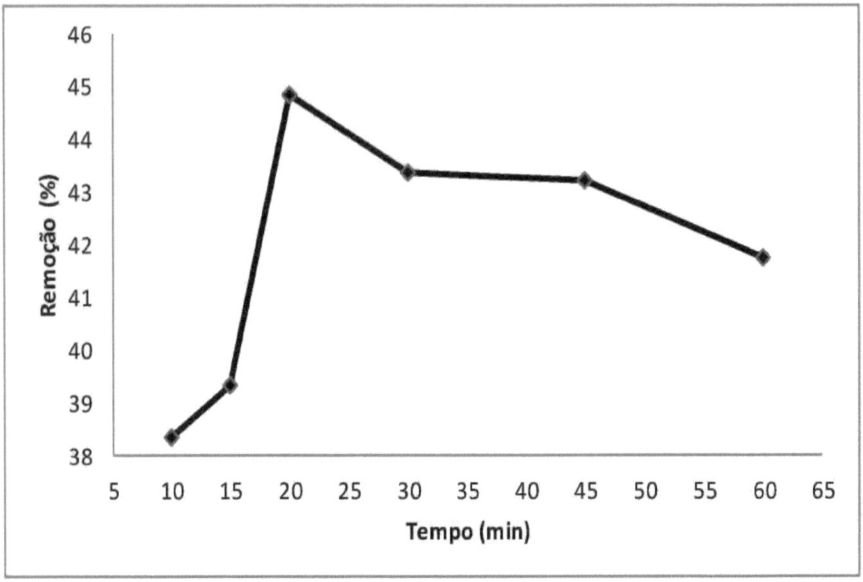


Conditions: m = 25.0 mg, volume of solution = 15.0 mL and [Cu (II)] = 10 mg $L^{-1}$ , pH = 7.0.

### 5.1.3 Study of adsorbent mass variation

For the same concentration of Cu (II) ions, the adsorption study was carried out as a function of varying the mass of crushed Baru bark and pulp. The amounts of mass used for the study were 25, 50, 100, 150 and 200.0 mg, as shown in Figure 10. The results show an increase in ion removal as the mass of Baru increases, with a significant increase at 200 mg. The experimental conditions used for the study were: volume 15.0 mL; stirring time 20 minutes; pH 7.0 and concentration of Cu (II) ions in the solution equal to 10 mg $L^{-1}$ .

In view of the results obtained, the mass chosen was 25.0 mg for the adsorbent of interest, because using smaller quantities of adsorbent guarantees a low cost for the procedure, reduces the influence on environmental impact and facilitates the experimental procedure by providing agility and speed. Furthermore, the rate of ion removal is not equally proportional to the increase in adsorbent mass.

Figure 10 - Effect of mass variation as a function of Cu (II) ion adsorption using Baru

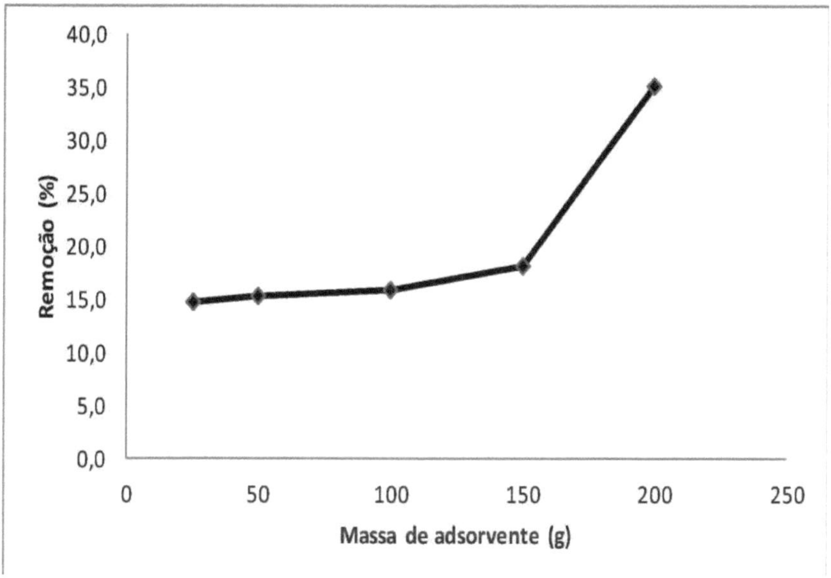


Conditions: pH = 7.0, time = 20 min, solution volume = 15.0 mL and [Cu (II)] = 10 mg $L^{-1}$

### 5.1.4 Maximum Adsorption Capacity (MAC)

Adsorption equilibrium studies to determine the maximum capacity for Cu (II) removal by Baru were carried out using adsorption isotherm models. The models used were those of Langmuir and Freundlich, which describe the equilibrium established between the metal ions adsorbed by the adsorbent ($Q_e$) and the ions remaining in the solution ($C_f$), at a constant temperature.

The isotherms are characterized by values that represent the surface properties and the affinity of the adsorbate for the biosorbent. For the study, the initial concentrations of the ions varied between 5.0 and 70.0 mg $L^{-1}$ of Cu (II), the mass of the adsorbent was kept constant at 25 mg, the pH of the solution was equal to 7.0 and the stirring time was 20 minutes, according to the optimization. Figure 11 shows the adsorption isotherm of Cu (II) by Baru.

The mass of the metal adsorbed was calculated by subtracting the value of the concentration of the working solution before and after the removal procedure and multiplying the value obtained by the volume of solution (L), dividing by the mass of the adsorbent (g) represented by the equation:

$$Qe = \frac{([S_i] - [S_f]) * V_s}{m}$$

Where: [Si] and [Sf] are the initial and equilibrium concentrations in $mgL^{-1}$ respectively, Vs is the volume of the solution (0.015 L) and m is the mass of the adsorbent (0.025 g).

Figure 11 - Adsorption isotherm of Cu (II) ions on crushed baru bark and pulp

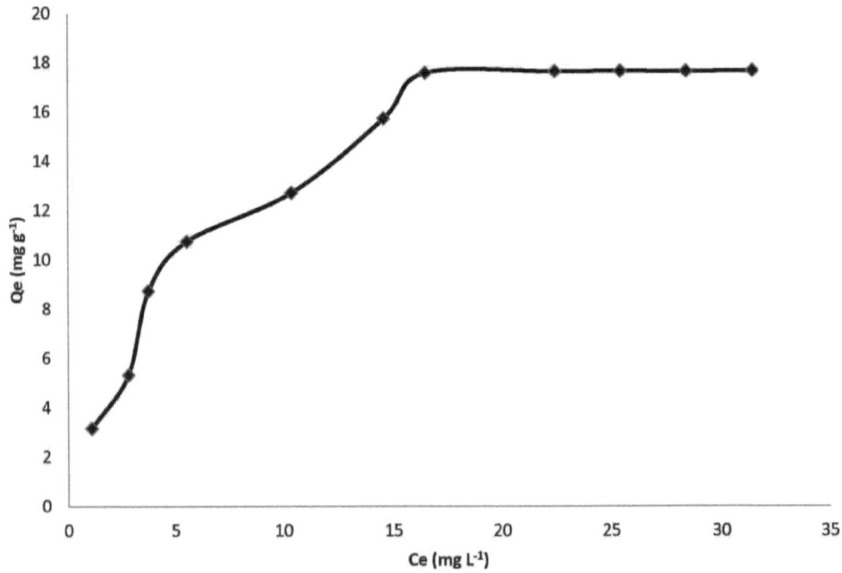


According to the classification of GILLES et al. (1960), the adsorption isotherm for the removal of Cu (II) ions resembles the L type. The initial shape of the L curves indicates that, as the available sites are filled, it becomes increasingly difficult for the adsorbate species to find an empty site. However, this convex isotherm is favorable because a high adsorption capacity can be obtained at low solute concentrations.

The experimental data shown in Figure 11 were applied to the Langmuir and Freundlich models, using the linearized equations of the models. The Langmuir and Freundlich constants obtained from the isotherm and the correlation coefficients are shown in Table 3 and the linearized Langmuir and Freundlich isotherms for the metal are shown in Figure 12 and Figure 13, respectively.

Figura 12 - Linearization of the Langmuir isotherm for Cu (II) using Baru.

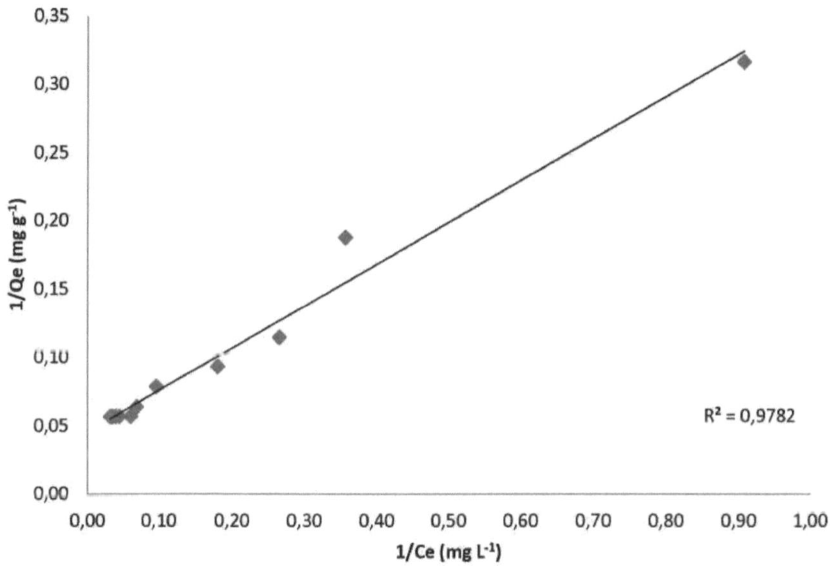


Figura 13 - Linearization of the Freundlich isotherm for Cu (II) using Baru.

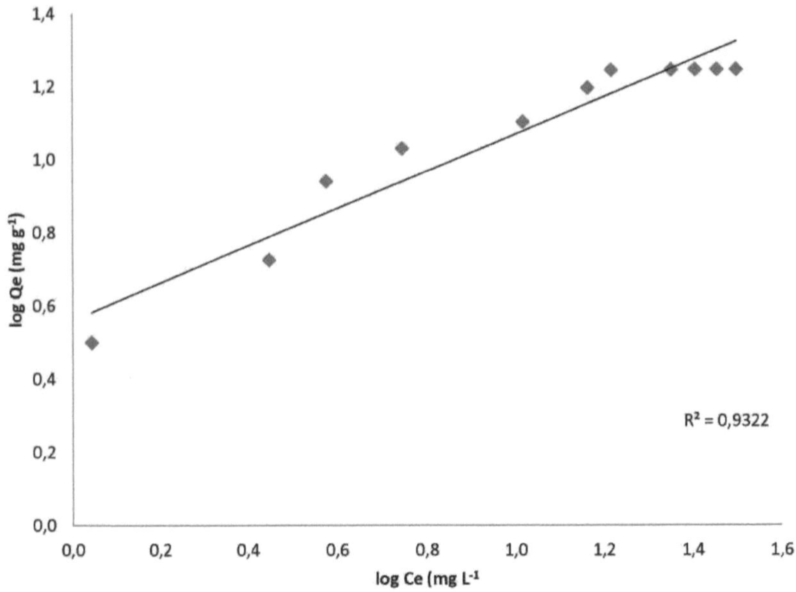


Table 3 shows the Langmuir and Freundlich constants and the correlation coefficient values from the isotherm for comparison purposes and to determine the best $R^2$ .

Table 3 - Parameters of the adsorption isotherm applied to the Langmuir and Freundlich model

| Langmuir | | | Freundlich | | |
|---|---|---|---|---|---|
| Qmax (mg g )$^{-1}$ | B | $R^2$ | *Kf* | n | $R^2$ |
| 22,12 | 0,1471 | 0,9782 | 3,645 | 1,966 | 0,9322 |

The values in Table 3 show that the Langmuir model fits the experimental data obtained for Cu(II), with a linear coefficient of 0.9782. The Qmàx value of 22.12 mg $g^{-1}$ obtained from the Iinearized equation of the experimental data in relation to the Langmuir equation, indicates a good potential for the crushed Baru bark and pulp to remove Cu (II) ions from aqueous systems. The $R_L$ parameter, Equation 3, shows that adsorption is favorable ($R_L = 0.317$) for the material, according to Table 2.

The Freundlich model showed a good fit with the experimental data, with a linear coefficient of 0.9322. The values of the Freundlich constants show favorable adsorption of Cu (II) by the adsorbent. The value of 1/n, which refers to the distribution of ions attached to the active sites, is represented by a value of 0.507, with values between 0 and 1 indicating that the ions of the species present are favorably adsorbed under the conditions tested.

## 5.2 Material characterization

### 5.2.1Infrared Spectroscopy (FTIR)

Infrared spectroscopy provides evidence of the presence of various functional groups in the structure of organic and inorganic compounds due to the interaction of molecules or atoms with electromagnetic radiation in a process of molecular vibration. Absorption in the infrared region is caused by rotational and vibrational movements of the molecular groups and chemical bonds of a molecule. Infrared radiation causes atoms in organic and inorganic compounds to vibrate with increased amplitude around the covalent bonds that connect them, giving rise to the bands observed in the spectrum (CUNHA et al, 2015).

To characterize the material, analyses were carried out on the "in natura" material after selecting the particle size and also on the adsorbent after adsorption and interaction with the analyte under study. Figure 14 shows a comparison of the results obtained:

Figura 14 - Infrared spectroscopy of Baru "in natura" (black) and
after adsorption (red)

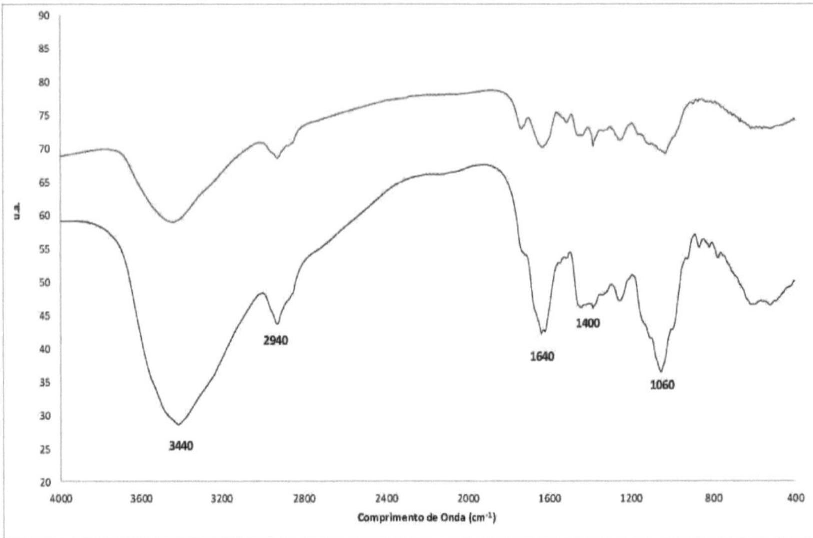


A comparison of the results shows that the bands of the "in natura" material were reduced approximately in the wavelength ranges of 3440 $cm^{-1}$ , 2940 $cm^{-1}$ , 1640 $cm^{-1}$ , 1400 $cm^{-1}$ and 1060 $cm^{-1}$ .

The broad band centered at 3440 cm-1 can be attributed to the stretching of the O-H bond and the stretching of the N-H bond of amides, due to the presence of proteins, fatty acids, carbohydrates and the lignin units present in baru bark and pulp. The band centered at 2940 $cm^{-1}$ corresponds to the stretching of the C-H bond of the primary and secondary carbons of the $CH_3$ , $CH_2$ groups.

The peak at approximately 1640 $cm^{-1}$ with a usually weak to medium intensity can be related to the bands in symmetrical alkenes. The band at 1400 $cm^{-1}$ can be related to the stretching of the bonds of the NH $group_2$ . And the region at 1060 $cm^{-1}$ is related to the stretching of the S=O bond.

The reduction of the respective bands in the material after contact with the solution may have occurred due to the interaction of Cu (II) with the adsorption sites of the material corresponding to the highlighted bands.

### 5.2.2 Thermogravimetric analysis (TGA)

Thermogravimetric analysis makes it possible to measure the change in mass of a material as a result of a physical or chemical transformation when it is subjected to a controlled temperature program where decomposition, crystalline change and sintering processes begin (DALPASQUALE, LOPES, ANAISSI, 2013). From the thermogravimetric analysis (TGA) curve and its derivation, it can be seen that between 25°C and 900°C, there is a loss of mass of around 73%, due to the elimination of water adsorbed by the material, decomposition of organic matter and volatile substances present.

The curve representing the loss of mass during the pyrolysis of the materials can be clearly divided into four stages relating to the most significant thermal events, which are related to different constituents. Between 25°C and approximately 140°C, there is a loss of mass of around 5.2%, associated with the loss of moisture, representing that the material had low humidity, probably due to the adsorbent preparation stage, where it was heated for 48 hours to eliminate excess moisture.

The second event, between temperatures of approximately 140°C and 198°C, with a mass loss of 15.8%, is probably due to the decomposition of the organic matter that makes up the protein fraction. The third and fourth events, between the temperatures of 198°C and 400°C, respectively, correspond to thermal destruction due to the loss of mass due to the decomposition of the three main components of the natural material, cellulose, hemicellulose and lignin, with a total loss of 45% in mass (SALLEH et al, 2015). Figure 15 and 16 show the results obtained from this analysis.

Figura 15 - Thermogravimetric analysis (TGA) of "in natura" Baru peel and pulp

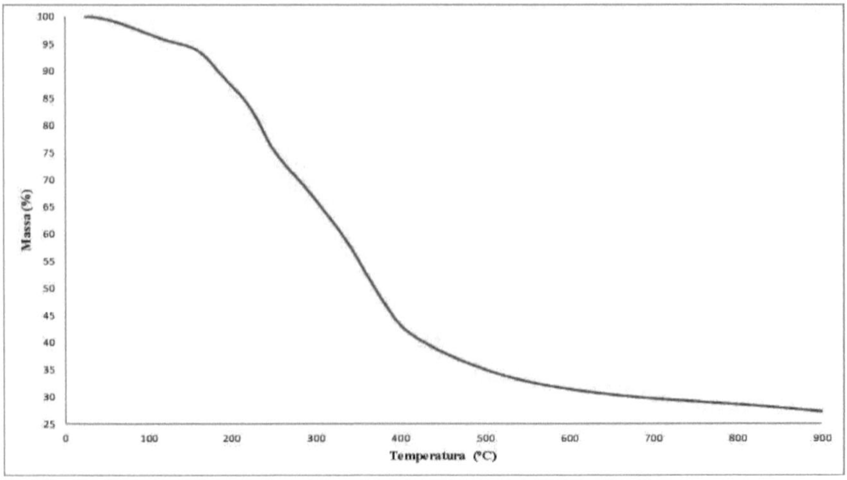


Conditions: adsorbent mass: 7.047 mg; heat flow of 10°C/min; Ti = 25°C and TF = 900°C

The TGA curve shows that at the end of the analysis, at a temperature of 900°C, the material still had approximately 27% of its initial mass, which is probably made up of ash.

Figura 16 - Derived from the Thermogravimetric Analysis (DGA) of Baru

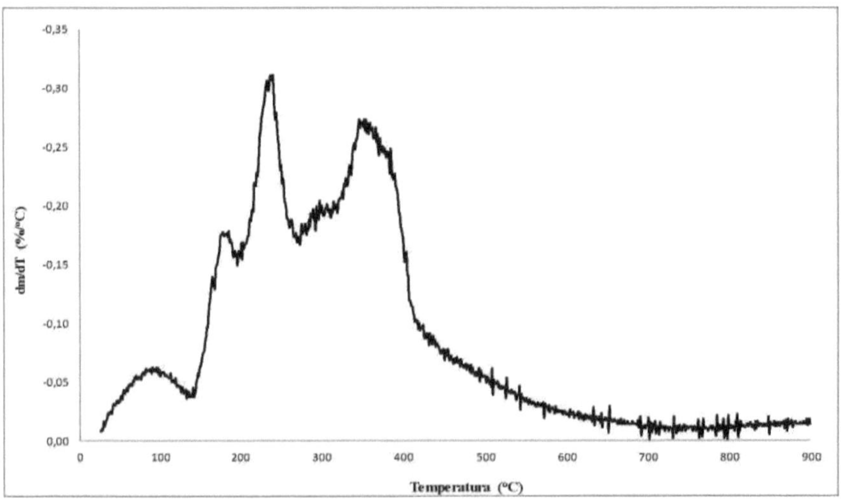


Conditions: adsorbent mass: 7.047 mg; heat flow of 10°C/min; Ti = 25°C and TF = 900°C

# 6 CONCLUSIONS

The removal of Cu (II) ions using Baru as an adsorbent showed an optimum pH of 7.0, a stirring time of 20 minutes and an adsorbent mass of 25 mg.

Applying the data obtained experimentally to the Langmuir and Freundlich models showed that the isotherms obtained were satisfactory for both models, with linear coefficients close to 1. The Langmuir model showed $Q_{max}$ equal to 22.12 mg $g^{-1}$ and $R^2$ equal to 0.9782 and the Freundlich model *Kf* equal to 3.645 mg $g^{-1}$ and $R^2$ equal to 0.9322. This suggests the occurrence of monolayer adsorption and also interaction between adjacent adsorbate molecules, due to the heterogeneity of the material.

The spectra obtained in the FTIR analysis of the material "in natura" and after adsorption showed the main functional groups that may be involved in the adsorption phenomena of the material under study. The technique made it possible to identify the presence of O-H, N-H, C-H, C=C and S=O bonds, functional groups which represent potential active sites for removing metal ions, which by electrostatic interaction enable adsorption through complexation or ion exchange mechanisms, as observed in the spectrum of the adsorbent after adsorption.

TGA showed the presence of water molecules, proteins, cellulose, hemicellulose and lignin in the composition of the material, by varying the percentage of adsorbent mass in relation to the gradual variation in temperature.

Analyzing the results obtained in the tests when using Baru *Dipteryx alata* as an adsorbent, they constitute a potential alternative for the removal of Cu (II) ions, since the proposed procedure is easy to carry out and can be used for additional removal of trace concentrations of Cu (II) for the polishing of contaminated effluents before discharge into water bodies.

# 7 REFERENCES

National Water Agency. Special section on the water crisis. **Conjuncture of Water Resources in Brazil**. 2014.

AMALINEI, R. L. M. et al. Investigations on the feasibility of Romanian pine bark wastes conversion into a value-added sorbent for Cu (II) and Zn (II) ions. **BioResources**. North Carolina: North Carolina State University, v. 7, n. 1, p. 148160, 2012.

AMAVISCA, C. V. Use of agricultural waste as a base for the preparation of activated carbon. Federal University of Rio Grande do Sul, Porto Alegre, 2012.

AYDIN, H; BULUT, Y; YERLIKAYA, Ç. Removal of copper (II) from aqueous solution by adsorption onto low-cost adosrbents. **Journal of Environmental Management.** Turkey, v. 87, p. 037-045, jan. 2008.

BACCI, D. C; PATACA, E. M. Education for water. **Institute for Advanced Studies**. Sao Paulo, v. 22, n. 63. 2008.

BIANCHI, M. A. The drinking water distribution problem for the city of Chapecó - Santa Catarina. Chapecó Regional Community University. Chapecó, Santa Catarina, 2011.

BRAZIL. CONAMA Resolution No. 430. Provides for conditions, parameters, standards and guidelines for managing the discharge of effluents into receiving bodies of water, partially amending and complementing Resolution No. 357, of March 17, 2005, of the National Environment Council (CONAMA). Ministry of the Environment, National Environment Council. Brasilia, 2011.

BRAZIL. SUSTAINABLE CONSUMPTION: Education manual. Brasilia: Consumers International/MMA/MEC/IDEC, p. 025-036, 2005.

CHAMBER OF DEPUTIES. Agenda 21. United Nations Conference on Environment and Development. Documentation and Information Center: Publications Coordination. Brasilia, 1995.

CARRAZZA, L. R; D'ÂVILA, J. C. C. Manual Tecnològico de Aproveitamento Integral do Fruto do Baru *(Dipteryx alata)*. ISPN, Brasilia, 2010.

COONEY, D.O. Adsorption Design for Wastewater Treatment. CRC Press, Boca Raton, Florida, 1999.

COSTA, L. A. V; IGNACIO, R. P. Relaçôes de Consumo x Meio Ambiente: Em busca do Desenvolvimento Sustentàvel. **Àmbito Juridico**, Rio Grande, v. XIV, n. 95, dec 2011.

CRUZ, K. S. et al. Partial characterization of proteins from Baru (*Dipteryx alata*) seeds. **Journal of Agricultural and Food Chemistry**, v. 91, p. 2006-2012, mar. 2011.

CUNHA, D. A. S. et al. Applications of Fourier transform infrared spectroscopy (FTIR) to characterize complexes. **ENEPEX.** Mato Grosso do Sul, 2015.

DALPASQUALE, M; LOPES, µ. C; ANAISSI, F. J. Thermal Analysis Applied to the Determination of Activation Energy. An Experiment for the Physics-Chemistry Teaching Laboratory. **Orbital: The Electronic Journal of Chemistry,** v. 5, n. 4, 2013.

DO, D. D. Adsorption Analysis: Equilibrium and Kinetics. **Series on Chemical Engineering**, v. 2, p. 913, 1998.

ELUMA S/A. Antimicrobial properties of copper: a range of possibilities for the benefit of human beings. **Eluma S/A**, Sao Paulo, 2010.

FREUNDLICH, H. M. F. Over the adsorption in solution. **The Journal of Physical Chemitry**, v. 57, p. 385 - 470, 1906.

GILLES, C. H.; MACEWAN, T. H.; NAKHWA, S. N.; SMITH, D. Studies in adsorption. part XI. A System of classification of solution adsorption isotherms, and its use in diagnosis of adsorption mechanisms and in measurement of specific surface areas of Solids. **Journal Chemical Society**, v. 846, p. 1-21, 1960.

GAVRILESCU, M. Removal of Heavy Metals from the Environment by Biosorption. **Engineering In Life Sciences**, v. 4, n. 3, 2004.

HASSAN, A. F.; ABDEL-MOHSEN, A. M.; ELHADIBY, H. Adsorption of arsenic by activated carbon, calcium alginate and their composite beads. **International Journal of Biological Macromolecules**, n.68, p.125-130, 2014.

KABATA-PENDIAS, A.; PENDIAS H. Trace elements in soils and plants. London: CRC Press, 1992.

LINHARES, L. A et al. Application of the Langmuir and Freundlich models in the adsorption of cadmium and lead in different classes of Brazilian soils. **Revista Tecnològica**. Maringà, v. 17, p. 49-60, 2008.

MARCATTO, C. **Environmental education: concepts and principles.** Belo Horizonte: FEAM, 2002.

MCKAY, G. Use of adsorbents for the removal of pollutants from wastewaters. CRC Press, Boca Raton, Florida, 1996.

MENESES, J. M. et al. Treatment of biodiesel effluent using electrocoagulation/flotation: investigation of operational parameters. **Quimica Nova.** v. 35, n. 2, p.235-240, 2012.

MESQUITA, R. F. et al. Methodological proposal and reflections on sustainable development and environmental education in high school. **Interdisciplinary Journal:** UNNOVAFAPI University Center. Teresina, v. 7, n. 2, p. 165-172, 2014.

Ministry of the Environment. Environmental education for a sustainable Brazil: ProNEA, Legal & Normative Frameworks. Brasilia, 2014.

Ministry of the Environment. Brazilian Agenda 21: result of the national consultation / Commission for Sustainable Development Policies and National Agenda 21 (CPDS). 2 ed. Brasilia, 2004.

MÓDENES, A. N. et al. Removal of heavy metals Cd (II), Cu (II) and Zn (II) by the biosorption process using the macrophyte *Eicchornia crassipes*. **Revista Escola de Minas**. Ouro Preto, v. 93, n. 3, p. 355-362, 2013.

MOSQUETTA, S. et al. Use of Baru *almonds* (*Dypterix alata*) for Ni(II) removal in fuel ethanol. **Quimica Nova**, v. 34, n. 6, p. 923-927, mar. 2011.

MUDHOO, A.; GARG, V.; WANG, S.. Removal of heavy metals by biosorption. **Environmental Chemistry Letters**. v.10, n.2, p.109-117, jun. 2012.

NAGATA, M. et al.. Sustainable development, corporate social responsibility, is it possible to be sustainable? The case of MAPFRE S.A. **Rumo Sustentâvel.** 2010.

PARAB, H; Sudersanan, M. Engineering a lignocellulosic biosorbent - Coir pith for removal of cesium from aqueous solutions: Equilibrium and kinetic studies. **Water Research**, v. 44, p. 854-860, 2010.

PEDROTTI, M; MISTURA, C. M. Avaliação de aspectos e impactos ambientais do processo produtivo de uma metalùrgica de pequeno porte. **Journal of Applied Exact and Technological Sciences.** Rio Grande do Sul: Universidade de Passo Fundo, v. 2, n. 2, p. 022-045, 2010.

PROSAB. Rational use of water and energy: Conserving water and energy in building and public water supply systems. **ABES**. Rio de Janeiro, 2009.

RIBEIRO, E. V. et al. Heavy metals and water quality of the sao Francisco river in the segment between três marias and perapora - MG: contamination index. **Geonomos**. Minas Gerais: Professor Manoel Teixeira da Costa Research Center, v. 20, n. 1, p. 049-063, May 2012.

RODRIGUES, M. A; SILVA, P. F; GUERRA, W. Chemical Element: Copper. **Quimica Nova na Escola.** v. 34. n. 3, p. 161-162, 2012.

RUTHVEN, D. M. Principles of adsorption and adsorption processes. John Wiley & Sons, New York, p. 432, 1984.

SABEI, T. R; BASSETTI, F. J. Eco-efficient alternatives for effluent treatment in rural communities. **Electronic Journal "Fórum Ambiental da Alta Paulista"**. v. 09: Health, Sanitation and Environment, n. 11, 2013.

SALLEH, M. N. et al. Influence of Ethylene-Glycidyl Methacrylate (EGMA) on TGA-FTIR Analysis of Recycled Polymer Blends and High Loading Biocomposite. **Journal of Advanced Research in Materials Sciences**, Malaysian, v. 9, n.1. p. 1321, 2015.

SANTOS, C. P. et al. Paper: how is it made? **Quimica Nova na Escola**, n. 14, 2001.

SEMEDO, M. F. F. R. Medico-legal importance of essential metals: copper and zinc. University of Porto, 2014.

SILVA, J. L. B. C. et al. Biosorption of heavy metals: a review. **Revista Saùde e Ciência.** v. 3, p. 137-149, 2014.

SINEVA, A. Adsorption of Synthetic Surfactants from Aqueous Solutions on Natural Adsorbents. In: The Role of Colloidal Systems in Environmental Protection. 1st ed. Moscow: Elsevier, 2014. Chapter 6, p. 143-170.

SOUZA, C. S. Chemical derivatization for the characterization of a lignin from sugarcane bagasse. Federal University of Uberlândia, 2006.

STEFFEN, G. P. K; STEFFEN, R. B; ANTONIOLLI, Z. I. Contamination of soil and water by the use of pesticides. **Revista TECNO-LÓGICA**. Santa Maria, 2011.

TAGLIAFERRO, G.V. el al. Adsorption of lead, cadmium and silver on hydrated niobium (V) oxide prepared by the homogeneous solution precipitation method. **Quimica Nova**, v. 34, n. 1, p. 101-105, 2011.

TERA AMBIENTAL. Guide to effluent disposal and treatment. 2014.

THEODORO, P. S. Use of electrocoagulation in the treatment of galvanic industry effluents. State University of Western Paranà. Paranà, 2010.

THEODORO, P. S. et al. Application of electrocoagulation in the treatment of galvanic industry effluents for copper removal. **CESUMAR:** International Meeting of Scientific Production. Paranà, 2011.

UCUN, H; AKSAKAL, O; YILDIZ, E. Copper(II) and zinc(II) biosorption on *Pinus sylvestris* L. **Journal of Hazardous Materials**, v. 161, p. 1040-1045, 2009.

©UNESCO. United Nations Decade of Education for Sustainable Development, 2005-2014: final document of the international implementation scheme. Brasilia: UNESCO, 2005.

MIX
Papier aus verantwortungsvollen Quellen
Paper from responsible sources
FSC® C105338

Printed by Books on Demand GmbH, Norderstedt / Germany